MATH
A FOUR LETTER WORD!

THE MATH ANXIETY HANDBOOK

Angela Sembera and Michael Hovis
with illustrations by
Matthew Hovis

**SPECIAL SECTIONS FOR
TEACHERS AND PARENTS**

Library of Congress Catalog Card Number: 90-90260

ISBN 0-9627036-0-5

Printed in the United States of America

I HATE MATH!

An Introduction

Of course you hate math. In fact, most people do. But there's a real world out there. And much of that world thinks math is pretty important. And really, deep down, so do you. I mean, you are reading this, aren't you?

This handbook was written for all of you who dread math, even if you are not in school. What we have to say is pretty simple. And it's all right here in this handbook that you can read in less than an hour. But you **do** have to read it. And then **you have to think about it** and **decide for yourself** if you want to do what it takes to do well in math. For you adults out there, it may even mean going back to school for a little while.

If you **do** make a commitment to approach math in a new way, be prepared for some hard work. You've probably got some pretty bad habits about how you deal with math and those are going to have to change. But we've made a career of teaching people who *really* hate math to overcome those feelings. And I promise that, if you follow some simple steps and invest some of your time, you can do well in math.

Who knows? You might even learn to like it. Really.

*This book is dedicated to Lynette
and all the other people who had the guts
to return to school, face a math class, and find
out how smart they really are.*

CONTENTS

PEOPLE WHO HATE MATH

Three Brief Biographies

*"I hate math! I don't need it.
I don't want it. I don't like it at all."*

MATT

The Math Hater

Matt doesn't just dislike math. He **hates** it. And he's sure that math is not for him. After all, *"That's why we have accountants and calculators."* Matt's hacked off that he has to take math to graduate and waits until the bitter end to take the math he needs to get his diploma.

Matt also hates going to class and feels that studying is just a waste of time. When math tests are failed, excuses are made, and offers for help are usually rejected. Matt thinks that if he **really** wanted to learn math he could, but *"Why bother?"*, since he has no intention of ever using it anyway.

OBSERVATION: People like Matt spend a lot of time and energy being angry and trying to avoid learning math. And this anger interferes with the concentration needed for learning. People who have a problem with math often convince themselves that it is not really necessary. Instead of admitting that they must work hard at math, they make excuses for why they don't need it. Math intimidates them. The thought of really trying to succeed in math and then failing is scary, so they don't give a full effort. In other words, **they cannot fail if they do not try.**

"*It's the family joke that I can't do math. When I have to use math to figure out even simple problems at home or at work, it's EMBARRASSING. I don't even know where to start.*"

JULIE

The Family Joke

People like Julie are usually female. Often, they claim to be just like their mother who *"wasn't any good at math."* She and others may joke about her mathematical incompetence. It's almost "kind of cute" to have trouble with math. *"I can't help it. That's just the way it is."*

OBSERVATION: Julie probably stopped taking math as soon as it became an optional subject. The thought of having to take math may even keep her from going to college. Usually she feels intellectually inferior to the men in her family, who are mathematically competent. If she is a parent, she is frustrated because she can't help her children with their math homework. Making change and balancing checkbooks make her nervous. The joking aside, the insecurity she feels from her mathematical experiences really does bother her, and her lack of math skills affects many of her daily activities.

"*When I have to take a math test, I feel physically ill. My palms get sweaty and my heart starts pounding. I get sick to my stomach and I feel like I'm going to throw up right there in class.*"

TONY

The Test Anxious Student

Tony is able to work problems correctly until it is time to take a test. Tests in other subjects are usually not a problem. But even the **thought** of a math test induces anxiety. Sometimes during a test, his mind will go blank or the problems will look unfamiliar. It is frustrating when the missed problems can easily be worked after the test is returned.

OBSERVATION: For students who expect a lot from themselves, the pressure to do well in school and make good grades is often overwhelming. And one bad experience can often lead to a cycle of fear. In other words, fear makes passing a math test almost impossible. And not passing the test increases this fear.

Do You Relate?

If you can relate to what people like Matt, Julie, and Tony feel and think, **you're not alone**. In fact, few people in our society really **like** math, and those that do are usually considered a bit odd. Hating math is contagious. It can be passed from one friend to another or from parent to child. In fact, most of us are **expected** to dislike math. And this expectation presents problems.....individually and as a society.

The U.S. is falling behind technologically. A strong technology requires a strong math foundation. Studies show that American students are not motivated to learn mathematics and, therefore, lag far behind their contemporaries in Japan, China, and the Soviet Union.

On an individual level, math avoidance hurts. Failing to understand mathematical probabilities can keep you from making good choices throughout your life. And not understanding numbers, statistics, geometry, and algebraic relationships will lead to poor decisions at home and on the job.

If you avoid math, you will not reach your full potential, and many personal and professional goals will be excluded. Math avoidance can even cause psychological damage by lowering self-esteem. In extreme cases, it can lead to stress and anxiety, which can make you physically ill.

CHAPTER 1

So Why Do People Hate Math?

The people examined in the above three cases all have the same underlying problem: a fear of failure. Their hatred, avoidance, and anxiety are symptoms of this fear.

SOMETHING HAPPENED: At some point, a negative experience with math occurred. It may have been a move or an illness, or perhaps a family trauma, such as a death or divorce. Or maybe it was a poor relationship with a teacher. In any event, something happened that caused the student to miss some basics. And math is a subject that builds on knowledge. So when some basic building blocks are lost, the student's math experience becomes repeatedly disappointing, and the student's view of math hardens into a narrow and dismal scenario of what never or always happens: *"I never pass"*......*"I always get confused"*......*"I always go blank"*......*"I never know where to start"*......

Anxiety, perpetual tardiness, indifference, boredom, poor conduct in the classroom, and endless excuses for not learning math are the result. **Emotions** have taken over the learning process. These responses to frustration lead to **"giving up"**. These students have chosen to stunt their own growth by not willing to try and not willing to risk.....**a classic fear of failure.**

What these people do not realize is that success usually does not come easily. Although Babe Ruth was the home run king, he was also a leader in striking out. But that didn't stop him from walking up to the plate. Abraham Lincoln failed in his first four attempts at elected office. But he, too, kept trying. Thomas Edison's saying, "Genius is 1% inspiration and 99% perspiration," really means trying and failing, and trying and failing, over and over again until success is achieved. **The path to success is usually paved with failures.** And successful people have learned to accept failure as a natural part of the growing process.

If we never fail, we are not putting ourselves into the **creative** situation of having to try new or alternate solutions to the problem at hand. Whether or not we fail is not what is important; it is what we do **after** we fail that counts. Learn to fail intelligently. When a mistake is made, we should not continue to make the same mistake in the same way. If we don't alter our strategies, then of course nothing will change. Mistakes are an indication that we must make adjustments and try again. Nothing succeeds like failure. Instead of running from failure, we should learn to embrace it and use it to our advantage.

"I swing big, I hit big or I miss big. And that's the way I live my life.....as big as I can."..............*Babe Ruth*

People who fear and hate math often rely on **three myths** that give them the perfect **excuses** they need to avoid it.

> *Myth 1:* *"We do not need to take math unless we are interested in a scientific or technical career."*
>
> FACT: **ALL CAREERS USE MATH.**

Employers often have to re-educate their employees to meet the demands of our more complex technological society. For example, more and more, we must be able to enter data into computers, read computer displays, and interpret results. These demands require math skills beyond simple arithmetic.

But the most important reason for learning math is that it teaches us how to think. Math is more than adding and subtracting, which can easily be done on a calculator; it teaches us how to organize thoughts, analyze information, and better understand the world around us.

SOME MATH PHILOSOPHY

The true nature of mathematics goes beyond a mechanical and impersonal experience. Unfortunately, there is not always time to leisurely explore the philosophical and creative side of mathematics in the classroom. But as we study and question mathematics, and really try to understand it in an unhurried environment, we begin to see how it is integrated into all facets of our lives. Math is related to **all** subjects, even English, music, drama, and art. So, math **is** important to our lives, even if we are not interested in a technical career.

> **Myth 2:** *"People who do well in math have mathematical minds, and usually these people are male."*
>
> **FACT:** *EVERYONE* **IS CAPABLE OF LEARNING MATH.**

There is no **type** of person for whom math comes easily. Even mathematicians and scientists spend a lot of time working on a single problem. Success in math is related to practice, patience, confidence in ability, and hard work.

For example, recent studies have shown that Asian students out-perform American students in math, not because they have "mathematical minds", but because their parents expect more of them and spend time helping and encouraging them. American parents (who are often uncomfortable with math themselves) do not have the same high expectations.

How many times do we or our parents feel relieved if we barely pass a math test? Do we feel the same way about an English test or any other subject that we feel good about? People learn to have low expectations when it comes to math.

The myth that only special people do well in math is extremely difficult to overcome. It is true that some people can solve problems or compute more quickly, but speed is not always a measure of understanding. Being "faster" is related to **more practice or experience.**

For example, the reason why math teachers can work problems quickly is because they've done them so many times before, not because they have "mathematical minds". And Larry Bird made a lot of free throws, but not because he's a "born natural". He would be the first to tell you it is because he had shot hundreds of thousands of free throws.

Working with something that is familiar is natural and easy. For example, when cooking from a recipe we have used many times before or playing a familiar game, we feel confident. We automatically know what we need to do and what to expect. Sometimes, we don't even need to think. But when using a recipe for the **first** time or playing a game for the **first** time, we must concentrate on each step. We double-check that we have done everything right, and even then we fret about the outcome. The same is true with math. When encountering problems for the very first time, **everyone must have patience** to understand the problem and work through it correctly.

> "Do not worry about your difficulties in mathematics; I can assure you that mine are still greater."..........*Albert Einstein*

STILL NOT CONVINCED? Remember.....the "myth of the mathematical mind" is extremely hard to overcome. People ask, "*Why do males score higher on standardized math tests than females? Doesn't this prove that some people have 'mathematical minds'?*"

There's a good answer for this. Until high school there are no real differences in mathematical performance between males and females. Later differences are a product of choices and expectations. In our society, many people, especially women, are "excused" from learning math. Studies show that men are not inherently more capable of learning math, but that **expectations** for women have not been as great. As children, females are often encouraged to be "good girls" and play with dolls, which involves talking skills; whereas boys are encouraged to "take risks" and play with mechanical toys that can be taken apart and put back together again.......activities which enhance mathematical abilities.

Not unexpectedly, American girls and minorities, who are often not encouraged or expected to continue with mathematics, fail to enroll in as many high school math courses as do white males. A study of Berkeley students reported that 57% of entering males, as opposed to only 8% of entering females, had four years of high school math. And a study of UCLA students found that 79% of its Asian students and 72% of its white students were in a precalculus math sequence. The figures for Hispanic and Black students were 25% and 20%, respectively. Of course white male and Asian students score higher on standardized math tests.....**they have had much more math!**

Math is not for the privileged few. And it is not magical. In fact, it can be learned by anyone. There is no secret to learning math. We may sometimes need help in understanding some of the language, but **with hard work, anyone can do well.**

Myth 3:   *"If we fall behind, or if it has been a long time since we've studied math, it's hopeless to try to learn it now."*

FACT: **IT'S NEVER TOO LATE TO LEARN.**

One of the main reasons people don't succeed in math is that they don't start at the right place. **IMPORTANT! You must begin where *you* need to begin.** Could you play a Mozart piece on the piano if you hadn't mastered "Mary Had a Little Lamb"? Why should learning math be any different?

If you change schools or if it has been a while since your last class, **you must determine what level math you should take.** A teacher or trained tutor can help determine this with a few placement tests and questions. So don't be afraid to **make sure** you enroll in the class that is right for you.

Sometimes a few tutoring sessions can help you fill gaps in your knowledge or help you remember some of the things you have simply forgotten. Or perhaps your foundations are weak and it would be better for you to relearn the basics. **Get some help** to determine what is best for you.

Taking the wrong course guarantees frustration. You will probably be too confused to even ask questions. But the real tragedy is that you may start believing that you are not capable of learning math. So **take the right course!**

NOTE: Sometimes you will be "required" to take a course for which you are not ready. And sometimes you may have to take a *review* course which covers several years of math very quickly. These situations are common for adults returning to school. If you think that your foundations are weak, **plan ahead**. You will probably need a lot of time for math, so you may want to take fewer courses or work fewer hours than you had originally planned. And before you register for classes, know when tutorial services are available. Then schedule your classes and work so that you will have time to get help.

THE MATH MONSTER

Don't forget. These **myths are just excuses** to avoid learning math. And the more we avoid doing anything, the more of a monster it becomes. Avoiding a fear leads to even more anxiety or self-doubt. For instance, people who have a fear of flying can overcome this fear only by flying. But if they schedule flights and then back out of them, their fear will become even greater. So don't let the **math monster** loom larger and larger. Confront your feelings and stop avoiding math.

CHAPTER 2

Understanding and Overcoming Cognitive Interference, Negative Self-Talk and Anxiety

Cognitive interference.....negative self-talk.....anxiety. These words help explain how we sometimes work against ourselves. An understanding of these words will help you combat your worst enemy..... **yourself!**

Many people have a fear of failing, and they let this fear interfere with their ability to learn. They are afraid of appearing foolish or of letting others see their flaws. They are oversensitive to public opinion. Often, they are perfectionists, who typically set high goals and worry about failure. Even though they are imaginative people, they generally underestimate their abilities.

These people feel more comfortable if they can memorize one "exact" way to work a problem or perform a task. They want to work in a simple, straightforward manner without having to take the risks involved in a learning process. When they do find themselves in a situation that requires taking a risk, they feel threatened, engage in negative self-talk, and become anxious and distracted.

HERE'S AN EXAMPLE OF THIS KIND OF THINK-ING: A student who is deathly afraid of math begins a math class. This student thinks, *"Why am I doing this? Math is so hard. I haven't taken it in years. I'm dumb. Everyone's staring at me. They know I can't understand this."* The student's mind wanders, *"I hope I remember* *to pick up my little brother.*" And then back to reality, as the student's heart begins to pound.....hands begin to sweat.....and stomach starts to ache.

In this example, the student is experiencing **cognitive interference, negative self-talk, and anxiety**.......the three demons. All three interact and feed on one another. They form a cycle that makes learning almost impossible. All three will produce bad feelings about math. So the more we know about them, the better we will be able to deal with them.

Cognitive interference is any thinking that interferes with what we are supposed to be doing. Sometimes it involves thinking about things that are completely unrelated to the task at hand, such as wondering about weekend plans.....thinking about errands.....daydreaming.....or worrying about a history test while doing math homework.

Learning mathematics is not simple. You need all of your brain working on the problems at hand. So if you find your mind drifting, **be aware**, and focus on the task before you. Now, I grant you that this can be very difficult. But the more you become involved in solving math problems and the better you become at the subject, the easier it will be for you to concentrate.

Negative self-talk is a type of cognitive interference. It is what we say to ourselves when we feel stuck, in trouble, or overwhelmed. In the example at the beginning of this chapter, the student lamented, *"I'm dumb. Everyone is staring at me. They know I can't understand this."* That, of course, is negative self-talk, and it is our worst enemy.

When we are confronted by difficulty, we often feel incompetent, which leads to a lack of confidence in our abilities. Instead of giving ourselves credit for being able to reach smaller goals, we let "the big picture" frighten us. For example, rather than working hard to understand one concept at a time, taking it one day at a time, math-anxious students worry about the whole course. And success at math, like anything else, does not happen all at once.

Sometimes negative self-talk is more difficult to recognize. Think of how many times, upon completing a problem or task, we say, *"That was too easy."* This remark implies that we are not taking credit for having learned to master a problem or do a job well. Instead, we should say something like, *"With practice, I knew I could do it ."*

Everyone engages in cognitive interference and negative self-talk or feels the symptoms of anxiety at different times.....during speech or drama presentations, sports tournaments, or dance or music recitals. For many people, anxiety occurs when they have to take a test, especially a math test. When we are nervous, it is not unusual to start thinking, *"I knew this was going to happen. I can't think. I feel like I've forgotten everything I've learned."* Statements like these make us feel even more nervous, and so the cycle continues.

NEGATIVE SELF-TALK

Negative self-talk holds us back. When it happens, we are letting our minds "play tricks" on us. We are setting ourselves up for guaranteed failure and it keeps us from taking the risks that must be taken to achieve success. When we expect to fail, our minds becomes receptive to failing. If we worry, *"What will others think?"* or *"What am I going to do if I can't pass?"* or *"What if I freeze and go blank?"*, then we will be much more likely to fail than someone who thinks positively.

These fears can also cause our bodies to release the hormone **epinephrine** as a reaction to the anxiety we feel. That's why we may feel jittery and sick to our stomachs when we are anxious.

OVERCOMING NEGATIVE SELF-TALK

Negative self-talk consists of **irrational statements** that must be recognized and changed. Keep track of your own negative self-talk. Write it down. Examine what you say to yourself. Learn to substitute neutral statements.

For example, suppose a student thinks, *"I'm never going to pass this test. I failed the last one, and I know the same thing will happen again."* Obviously, if a student has gotten help since the last test and is well prepared, failure is unlikely. The student is not taking credit for setting the goal of working hard and then doing so. A good rational statement to substitute might be, *"I've worked hard since the last test. I've learned from my mistakes, and I know that I understand the material, so I have a good chance of doing well on this test. Let's see how I do."*

When overcoming negative self-talk, it is not a good idea to make substitute statements that are **overly** positive, such as, *"I know that I am going to make an 'A' on this test."* Such statements can create a new threat and a new fear of failure. Again, substitute with **realistic, neutral** statements while remembering to give yourself credit for what you have accomplished.

Other examples for eliminating negative self-talk

"I just know I'm going to fail the test."
If I study, I bet I can pass.

"I'm afraid to answer questions in class, because everyone will laugh at me if I'm wrong."
Even if my idea is wrong, by discussing it, I can learn more about the problem.

"It's almost my turn to bat. I'll probably make the third out."
It's my turn to bat. I'm going to concentrate and really watch the ball.

"On stage, I'll be so scared that I'll probably forget my lines."
I've always wanted to act. I'm going to relax and enjoy the experience.

"I'll never be able to write a ten-page term paper."
I've got some good ideas and I know I can put them on paper. I just need to sit down and start.

"I'm just not capable of balancing my checkbook."
I know how to add and subtract. If I take the time to write down every check and every deposit, I can keep a balance.

Replacing negative statements with positive statements takes practice. But research has shown that people who learn to reduce their negative self-talk are able to improve their performance greatly.

Anxiety and fear are closely related. Fear is a survival mechanism. In life-threatening situations, such as an animal attack or car wreck, it produces chemical changes to give extra energy to ward off danger.

But with **anxiety,** the threat is **mental,** and there are no physical means to release the energy. So anyone who feels the intense physical symptoms of anxiety.....sweaty palms, nervous stomach, rapid heart rate, tense muscles, and so on.....cannot concentrate on the task at hand.

Like fear, anxiety produces adrenalin, which stimulates the heart, raises blood pressure, liberates sugar stored in the liver, and contracts groups of muscles, because a perceived threat or crisis is at hand.

With some math students, anxiety is extreme. They actually feel as if their lives are at stake. Often they make statements like *"I'll just die if I don't pass this math test."* These feelings are very real. They see their whole futures hinging on passing a math class.

OVERCOMING ANXIETY

Just being aware of your negative self-talk and cognitive interference is a good start. But there are many other positive steps you can take to overcome the **mental** aspects of anxiety.

- Acknowledge your feelings. Admit that you are anxious.

- Stop yourself from thinking irrelevant thoughts or putting yourself down.

- Rework your negative statements into neutral statements and think in positive terms.

- Don't worry about what others may be doing or thinking. It's what we do and what we say to **ourselves** that count the most. Think "*I can*" or "*I want*" instead of "*what if.*"

- Know that even failure has a bright side: you can learn from your mistakes. Remember, if you do not take risks, you are not growing. And taking risks means allowing yourself the **freedom** to fail.

- Don't worry about **everything** at once. Set goals that you can accomplish one step at a time. If you occasionally stray from your goal, don't give up on yourself. It's OK to feel guilty for a little while, but resolve to get back on track.

- And finally, **focus your attention away from yourself** and toward the task at hand.

NOTE: Keeping a notebook or diary of emotions, feelings, or thoughts can help you overcome anxiety or anger. **Any time** you feel frustrated, such as doing homework, studying for a test, taking a test, making change, or balancing your checkbook.....note the following in a journal:

1. What were you doing when you started feeling anxious or angry?
2. Where were you?
3. In general, how did you feel physically?
4. In general, how did you feel mentally?
5. **Exactly** what did you say to yourself?

Noticing the circumstances that triggered anxiety or anger will help you understand when it is likely to occur. Awareness of negative thoughts and feelings is the first step in gaining the necessary ammunition to combat the mental aspects of anxiety.

This mental approach will be helpful, but what can we do about the **physical** aspects? When we start feeling nervous or anxious, we obviously need to relax, but we cannot always talk ourselves into relaxing, especially when taking a test. Here are several techniques for overcoming the physical aspects of anxiety. And you know what? They really do work, so give them a chance.

- **Exercise.** A few hours before a test or any other situation which causes anxiety, engage in a **mild aerobic** exercise for 15 or 20 minutes. Anything that will increase your heart rate and get your blood moving, such as a jog or a brisk walk, will work. Exercise will help release some of the pent-up energy that comes with anxiety. You will also be more alert mentally because you have increased the flow of blood to your brain. Blood "feeds" the brain by supplying it with oxygen. Thus you will go into a situation that normally causes anxiety feeling more refreshed, alert, and relaxed.

- **Avoid caffeine.** Coffee, tea, and chocolate are stimulants and can make you even more anxious.

- **Practice muscle relaxation** 15 to 30 minutes before the anxious situation. Slowly and progressively tighten each group of muscles and then relax them. Start by taking a deep breath and feeling your chest muscles become tight and rigid. Then slowly exhale and think the word "calm." Feel yourself relax as you breathe out slowly. Repeat several times and move on, focusing in turn on your face, neck, shoulders, arms, hands, stomach, and legs.

- Practice **deep breathing**. This exercise is quick and especially helpful when taking a test. The steps are:

  1. **Instead of fleeing anxiety, face it.** Feel the anxiety for a few moments. Then, for a few seconds, put down your pencil or pen and sit back from the desk. Either close your eyes or look into space. Do not consciously focus on anything in particular. Try to clear your mind and relax your muscles.

  2. **Take three very slow, deep breaths.** Release each breath very slowly. Continue to let your mind drift. If you have trouble letting go of your thoughts, then imagine a scene or activity that is enjoyable.....a favorite hobby.....a walk along the beach with the waves gently lapping on the shore.....the sound of water rushing along in a river. Try to imagine the scene with all your senses.....hear it, feel it, see it, and smell it.

  3. **Slowly draw your attention back to your work.** Tell yourself that you are now relaxed and ready to perform the task at hand.

- Periodically sit up and **stretch** your **back and neck muscles** to prevent or alleviate tension and stress. There are several easy ways to stretch the back and neck without distracting the class. One way is to sit up straight, keeping your head forward, and slowly twist your shoulders from side to side. Another is to sit up, close your eyes, and slowly rotate your head from side to side.

YOU CAN DO IT!

You *can* break the cycle of cognitive interference, negative self-talk, and anxiety by positive thinking and relaxing. Sports psychology professes these two approaches as being key ingredients to winning. And just as you must practice math, you must also practice relaxing. Do not wait until a critical time, such as a test, to try deep breathing for the first time. Familiarize yourself with relaxation techniques. Anxiety won't necessarily go away, but over time and with practice, your ability to confront all of the anxiety in your life will greatly improve.

> **NOTE:** **Anger**, like anxiety, is a reaction to the frustration felt when there is an obstacle to reaching a goal. We hear, *"I just can't do it!,"* or *"That teacher can't explain anything!"*

As with anxiety, the first step is to acknowledge feelings. Only when angry feelings are ventilated can progress be made toward looking at the problem and finding solutions. Say to yourself.....or a teacher, friend, parent, or counselor....."I'm angry!" Then try to identify and understand the source of the anger. Remember, anger is an emotional response which is usually not rational. Examine whether or not your feelings seem out of proportion. Should you be as angry as you are? Don't fight fear with anger. **Take responsibility for what the problem really is and work to resolve it.**

> "You gain strength, courage, and confidence by every experience in which you really look fear in the face....You must do the thing you think you cannot do."........*Eleanor Roosevelt*

CHAPTER 3

How to Learn, Understand, and Succeed in Math

As with exercise or dieting, learning math will seem overwhelming at first, and just getting started may seem impossible. But when you *do* spend time exercising or you *do* stick to your diet (even when you don't want to) you feel great afterward, and with each passing day, you enjoy it more. The same is true with learning math. **But you have to sit down and do it.**

Many successful people will tell you that the key to their success is setting goals. And their desire and determination to reach those goals keeps them on track even when setbacks occur. "Failures" are looked upon as opportunities from which they can learn.....opportunities to do better or make progress toward their goals.

Attitude has everything in the world to do with success. First, **think of math as an associate, not an opponent**. Then it will be easier for you to set aside the time you need to study and practice problems. For high school and college students, at *least* an hour or two a day, in addition to class time, is necessary to ensure understanding, which is the key to doing well.

Don't be tempted to give up when you don't understand. This temptation to avoid math is strong and dangerous. There will be many times when you will not know how to work a problem. That is why you are taking the class.....because it is covering material that you need to learn. **The first step in a learning process is *not* under-**

standing. If you become frustrated because understanding does not come quickly, remember this is normal. It takes time.

The first step is always difficult. But as you master the techniques, you will find that you **can** do it. The same is true in music or sports. As you continue to work and practice, playing the scales or catching a ball becomes easier. (Remember when you were first learning to ride a bike, or roller skate, or tie your shoes?) We all have to work hard to become competent at a new task. It only seems easy after we know how to do it.

Here's a math example. When studying for a final, have you ever reviewed the material from the beginning of a semester and thought something like, "*That stuff seems so easy now. I can't believe I used to think it was so hard.*" ? Well, the reason it seems easy is because **you**

have learned it. It was difficult at the beginning because you were learning something new. It is important to know that learning does not happen in a smooth upward line. Learning occurs in stages or **plateaus**. For a time, learning will come easily and you will progress smoothly. Then suddenly you will hit a wall, and then you will struggle to reach that next step, or higher plateau of learning. So keep on trying, and remember that success happens one step at a time.

Question: *How do you eat an elephant?*

Answer: **One bite at a time!**

SCHOOL DAYS

OK. You've made a commitment to learn math. You have, haven't you?! Good! You know more about overcoming the psychological barriers, but now you have to decide how best to invest the many hours you will spend in the classroom and at home.

How to Get the Most from Your Class Time

- We talked about this earlier, but it is so important that we want to talk about it again. **Make sure that you are in a class that is right for you.** Talk with a tutor or counselor. Explain your past experiences with math. If you passed prior classes but did not really understand the material, take a placement test to see what you really know. Begin where *your* knowledge left off. Do not be afraid to review a class if a placement test shows you have gaps in your knowledge. Even if a review class seems "too easy" at first, starting on sure footing is better than falling behind in a class for which you are not prepared.

- Know that **you are capable of understanding the course.** As long as you have the prerequisites for the class, **you can** master the new material. You will need to work hard, and sometimes you may need help, but courses are designed so that students can pass. Your school will not offer a class that normal students can't pass. Really!

- Make sure that you are **on time** and always **bring a pencil, your textbook, and a notebook.** Arriving late distracts the teacher and class, and you run the risk of not understanding the entire lecture when you miss the beginning. Having the proper tools to do any job is a must. Borrowing paper or sharing a textbook in class is a nuisance for everyone. A pencil with a working eraser is extremely important when solving math problems. Scratching out errors is sloppy and makes reviewing notes or problems difficult.

- It is a good idea to **sit near the front.** Other students will be less likely to distract you, and you can see the board and hear the teacher more clearly. It is important that you devote all your attention and concentration to the lesson. The notes must be accurate, so seeing the board and hearing the teacher is critical. Think back to your earlier school years. Didn't it always seem that the best students sat at the front? Well, here's your big chance.

- **Take** lots of **notes.** Have a well-organized notebook with a special section for your class notes. If you are unsure of anything that your teacher writes or says, be sure to ask for more explanation, even after class. It is important to record everything correctly. Fully understanding the day's lesson may have to come later when you have more time to study your notes.

- **Do not let the appearance of math intimidate you.** Be prepared to learn and understand the **language** of math. Studying math is often like studying another language. How many times have you looked at a mathematical expression and said, *"It looks like Greek to me."?* In fact, it often is Greek. Mathematicians use Greek letters of the alphabet for many of their symbols, and these symbols have special but simple meanings which become perfectly clear after you learn them.

- **Do not compare** your progress with that of your classmates. Concentrate on your own progress. Everyone's mathematical background is different. Many of your classmates may be there just for review. These students have already learned the material and will not need to struggle for understanding. Don't let them, or the "class brains", intimidate you. Remember, your classmates do **not** have "mathematical minds". They have simply spent more time learning math.

- **Ask questions** in class. Most students who don't feel comfortable with math are afraid of asking questions in class. They say to themselves:

"I'm afraid the teacher will know how dumb I am."

"I feel like I'll hold up the whole class."

These students are so busy worrying about what everyone else thinks that they forget they **have the right** to ask questions. And they have the right to not understand. Assuming that you have been paying attention, the class will welcome your questions. Questions are never "dumb." They are simply an indication that there is a gap in your understanding. Most likely other people have the same problem and will be glad that you asked. As one student said,

"I would find myself in the middle of a math class not knowing what I was learning. When I asked the teacher, it seemed the whole class understood better."

Questions show that you are thinking and trying to understand. A teacher will often gain new insight from a question. Even if you don't know exactly what to ask and simply say, *"I just don't understand, can you explain it differently?"*, everyone, including the teacher, will learn something from the new explanation.

Often a teacher will skip steps when discussing a problem. If you are confused, let the teacher know. It is your responsibility to let the teacher know when to explain more fully. Sometimes teachers will assume that a lack of questions from the class implies that everyone understands. Not wanting to bore the class, they will continue to skip steps, unless someone speaks out. **You are doing everyone a favor by asking questions or stating that you are confused.**

MORE MATH PHILOSOPHY: Getting comfortable with math requires more than just memorizing formulas. If you really try to understand the material, you will often have many insights and questions that are very stimulating for you and your teacher. **Mathematics and philosophy are closely related.** Try to see math beyond a rigid set of rules to be memorized, and it will become much more interesting.

If you are just beginning college, here's something to think about: far too many students drop out or fail their very first semester. Here are three ideas that should help:

1. **Don't overload your class schedule.** As a rule of thumb, for every hour spent in class, you should study at least two hours outside of class. So if you enroll for five classes, which is usually 15 hours of class time, you are committing yourself to at least a 45-hour week for schoolwork (check my math). This accounting makes it nearly impossible for students to work and go to school full time. But you'd be surprised by the number of students who try it and fail. So, if you must work, take fewer classes than a full time student, at least for your first semester. Make sure you **get off to a good start** by having enough time to enjoy your educational experience.

2. **Take your math class at a time when you know that you will be alert.** If you have trouble with math, don't sign up for a 7:00 a.m. class when you know that you like to sleep late. Remember, you need all your brainpower and concentration for understanding math.

3. **Always include at least one class that you think you would really enjoy.** Read the course catalogue to get a description of courses that sound interesting to you. Better yet, talk to a teacher about the course and materials to be read to see if the class really is of interest. Learn to live a little.....give yourself a treat each semester, and you will be much more likely to appreciate and continue your education.

How to Study and Practice Math

Now this may *sound* like some bad news.....but if you want to do well in math, there is something to **work on every day.** Even if no problems are assigned, there is still material covered that you should review and practice from the textbook and your notes. Spending the time you need may be difficult at first, but when you attain a little success, working on your math can be a challenge you enjoy.

- **Establish a routine** for working and studying. Find a **place** to work that is comfortable and use this same place every day. Set aside a **time** when you are mentally alert and not likely to be bothered. Let your family or roommate and friends know that this is your time and place for working math. Do not let anyone or anything distract you. Be committed. Eliminate any urges to interrupt your math routine. Watch out for statements like *"I don't feel like doing my math now. I'll do it tomorrow."*

Establishing a familiar, daily routine will make your commitment easier. With each passing day, you will feel more comfortable and get to work more quickly. Your routine will serve as a stimulant and provide positive reinforcement for working on math. You will have created a relaxed atmosphere where learning can be fun.

- **Read the pages in the text** that correspond to the class lesson. First, read your text before class. Even if you do not fully understand everything, you will get much more out of the class lecture than students who did not read the material. Then read the pages again after class. Now the words will be more understandable and the lecture will be reinforced.

- **When you come to a definition or theorem, pay special attention.** They may often seem complicated, but they always state a basic idea that is crucial to the lesson being taught. Do not let the language of math mystify you.

  A lengthy definition or theorem has phrases which must be examined individually. If you don't understand some of the words in the phrases, look them up. After you know what the individual pieces are saying, it will be easier to understand the whole definition or theorem. Then write it down on a separate sheet of paper. Say it out loud. With some thinking time, you can grasp the concept. And if you just can't quite get it, get help until you do.

- When necessary, **use books other than your text**. Remember, there are many different ways of explaining the same concept. Often, other books will be important resources that can help you understand a math lesson more fully. Your teacher or librarian can help you find books appropriate for your level of learning. Use the index to find the concept you are studying.

- **Pay attention to the written instructions** for each problem. Instructions define the problem and contain key words that help you know what to do. Many problems can look exactly the same, but with different instructions, they are different problems.

  Students often begin homework without **reading** the instructions. Then, when taking a test, these students become confused because they see many familiar looking mathematical expressions, but they do not understand what is being asked. They have not learned how to follow instructions, which is the key to solving any math problem.

- **Understand formulas before you memorize them**. People who believe that they are incapable of learning math try to memorize formulas without understanding them. They hope that math can be learned as rigid sets of rules that will apply in the same way all the time. Then they often become confused on a test when problems aren't exactly like the ones they have practiced at home. In addition to understanding **what** a formula does, make sure you know **why** and **how** and **when** to use a formula.

- Make a **separate study sheet for each concept** you are learning. Include any definition or theorem that relates to the concept. Include an example problem. (Don't forget to write out the instructions.) Having a separate sheet of paper that focuses on only one concept will help eliminate confusion. It is also a good way to organize material and make formulas easier to learn. These study sheets will be your "flash cards" that can help you review for a test.

- **Practice problems.** Review the example problems both from class and the textbook. Then copy these problems down on a separate sheet of paper. Without looking at the book or your notes, try to work them yourself. You may have to struggle for a while, but that is part of learning. When you get stuck, compare your work to your teacher's or the book. Once you fully understand the examples, you will be ready to practice problems from the exercises.

Like learning to cook or play a sport or musical instrument, you must practice and then practice some more before you will be able to perform competently. Understanding is not enough. For instance, you might understand how baseball is played, but you would not be able to play well if you had never practiced. And the more you practice, the better you become. As Edison said, "There is no substitute for hard work."

Other countries use practice as a key to succeeding in math. For example, a Japanese method of individual instruction has each student work on one lesson until he or she is able to get the problems 100% correct. Striving for perfection has its benefits. Students are taught to have high expectations for what they can do, and they leave the lessons feeling good about what they have accomplished. And the practice pays off when it comes time to take a test. Everyone feels pressured under stress. But the more you practice, the more familiar the problems will be. Math, like sports, music, or dance, requires practice for a more effortless performance.

- When first learning how to work problems, **do not place too much emphasis on the correct answer**. You should be more concerned with whether or not the **steps** to working the problem are correct. Students who race to the back of the book to see the answer can slow the learning process.

Even if their incorrect answer is the result of a careless error, they probably feel lost. On the other hand, if their answer is right, they are often so relieved that they quickly go to the next problem without being sure the problem was **worked** correctly.

So do not rush to a solution. **Always write out the steps involved in solving a problem in a neat and organized fashion with one step below the other.** When you do get a wrong answer, do not immediately erase your work. First look for what you did right and what you did wrong so that you can learn from your mistakes.

- **Never sacrifice correctness and understanding for speed.** When you are first learning how to work a problem, you will want and need time to understand **why** you are learning the problem and **how** to break it into steps that you can understand. Rushing will make you frustrated and anxious. Learn to relax and to question and enjoy the learning process. Speed will come after you understand your math problems and have had lots of practice.

- When studying math, **say it and hear it as well as write it**. Work problems where you will not bother family members or classmates, and speak, sing, or chant each step aloud as you write. This may sound silly, but it works! For instance, many people say new phone numbers out loud to help remember them. By speaking and hearing, as well as writing, you use more parts of your brain. Different kinds of memory paths are developed, and learning is enhanced.

- **Put what you have learned in your own words**. Once you have studied, find someone to whom you can verbally explain the lesson. A family member or classmate is a good choice. Together you can talk about what the material means. It may seem awkward at first, but it will get easier. You will be surprised to see how much more you can learn by trying to verbalize your thoughts.

- Watch out for your **handwriting**. Take note of when your work becomes sloppy or your handwriting changes. Writing lighter, bigger, smaller, or messier indicates that your feelings are affecting your work. For instance, bearing down harder with your pencil can indicate anger or frustration; writing smaller or lighter might mean that you are unsure of yourself.

Being aware of such **feelings** is important. They influence your attitude, which influences your learning. Knowing how you feel when you work problems will help you understand what attitudes you need to control or change to make math a more positive experience. Do not let your feelings control you; learn how to control and use them to your advantage.

- **When you get stuck, don't quit.** By the same token, **don't go in circles** on the same problem. When you are confused or cannot correctly work a problem and you have spent some time trying to clear up the difficulty, get help! If help is not immediately available, then make notes to remind yourself to get help later. (Remember: Do not erase your work!) While waiting for help, move on. Sometimes, you **will** be able to work the next problem or understand the next section of material.

Suppose you've taken notes and tried reading the book and working problems, but you still do not understand. When this happens, **get help from others**.

If you think that only *dumb* people need help with math, then you haven't overcome the myth about mathematical minds. **Everyone** can use all the help they can get when trying to understand something for the first time. So line up your resources at the beginning of a semester.

- Your **teacher** is the first person you should see for extra help. Make sure you know when he or she is available. In addition, he or she can refer you to other programs or resources. Your teacher will appreciate your wanting to understand and do well in math. And every time a teacher helps a student, he or she learns something about that student or how to better explain a mathematical idea.

- Ask about any **tutorial programs or math labs** your school offers. Many times they are free, but the hours may be limited. Arrange your schedule so you can take advantage of this help.

- Sometimes **a family member** can help. Let him or her know if you are having trouble with your homework. Maybe together you can figure out your problem. At the very least, your parents need to know right away if you do not understand, so they can help you find help before you get too far behind.

  If someone in your family does help with your math, be careful. Sometimes family members are impatient, which can lead to frustration or even shouting. When this occurs, it is better to get help outside the home.

- **Friends, especially classmates**, are an excellent source of help. A recent study of calculus students at a California university discovered that the main reason Asian students do so much better than other students is that they spend a lot of time studying together.

Studying or doing homework together will make learning math much more enjoyable. Explaining problems to each other increases understanding. Making a date to study together will decrease the chances of your avoiding math and will assure that everyone gets the job done. However, you do need to be careful. People often have different ways of working the same problem. As Emerson said, **"Each mind has its own method."** If someone works a problem in a different way from you, that doesn't necessarily mean one of you is wrong.

- **Seek private tutors**. Many times a school or school district will have a list of math tutors you can call. The prices they charge and their level of experience will vary. College students can make good math tutors and they usually charge less, but it is hard to beat the experience a veteran tutor can offer. Spend some time on the phone with tutors to pick the one that is best for your needs. Find out their math specialty. Ask which age group they most often tutor. Don't be shy about asking for references. It is important that you get the best tutor for you. Feeling comfortable with your tutor is important. A relationship of trust will make it easier for you to ask questions or let him or her know if you still don't understand.

Educational studies show that one-on-one teaching is the best way to learn. So if your school offers this service or if you can afford an individual private tutor, you will have an excellent opportunity to greatly increase your understanding of math. If you cannot afford private tutors, **get creative.** Contact local social organizations, such as your church or the United Way, for referrals to volunteer tutors. And remember, tutoring help is usually short term.

No matter where you get help, there are five important things to remember:

1. If after reviewing all the material, you still don't understand, **get help right away. Do not wait until you get too far behind** and are feeling really helpless. If you make sure you **understand** math, you will like it and do well.

2. **Be prepared before you ask for help.** Read the material, go over class notes, look at examples, and try to work the problems.

3. If you get stuck on a problem or cannot arrive at the answer, **do not erase your work**! Looking at what you've done wrong as well as what you've done right will help the tutor understand where you are having trouble.

4. **Ask questions!** Teachers or tutors are paid to help you, and if you do not ask questions, they will not know how to help. Even if you're not sure exactly what to ask, you can still say you don't understand. If after an explanation you still don't get it, say so and ask if they can explain it another way. Whatever you do, **don't pretend to understand**. Have your helper back up, if necessary, or explain the concept in different ways until you are confident that you can work the problem.

5. Don't be afraid to **let the person helping you know how you feel about math.** Share your math history with your teacher, or friend, or tutor. Let them know about any bad or good experiences you may have had. Let them know what your fears are. Let them know what helps or hurts. You will be surprised at how much more people can help if they understand from where you are coming.

CHAPTER 4

How to Reduce Math Test Anxiety

People who fear math feel the most anxious when confronted with a math test. In addition to the methods discussed earlier, several other positive steps will help reduce test anxiety.

BEFORE THE TEST

- The unknown or unexpected is usually what you most fear when taking a test. You can eliminate this source of fear by knowing exactly what material the test will cover. If not written or announced by the teacher, ask! **Ask what pages to review and what kind of problems you need to know.**

  But don't nitpick by saying, "Do we really need to know how to work these problems?" You need to know **all** the material, unless told otherwise. Do not ignore the ones you don't like or find difficult. They are not going to go away. Get help with these problems. **They wouldn't be taught if you could not do them.** Once you understand difficult problems, you won't mind them, and you will be a lot less nervous going into the test knowing that you have adequately prepared yourself for **all** types of problems.

- **Don't cram** for math tests. If you do, then you will be unsure of yourself while taking a test. **Begin your review several days in advance.** Dig out all those big study sheets you made and make sure you understand the principles they teach. Then work on practice tests. They can be chapter tests in your textbook, or you can ask your teacher or tutor to find or make up some for you.

  You can also make up your own practice tests. On a separate sheet of paper, copy a couple of problems from each exercise you have studied. Be sure to write out the instructions for working each type of problem. Close your notes and textbook and pretend these problems are a test.

- The **night before** the test, you should just review what you have already studied. **Do not stay up late.** Remember, you began preparing for the test several days earlier. And if you don't get a fair amount of rest and sleep, you will defeat all your good intentions. Budget your time that day so you can complete your review and still go to bed at a decent hour.

- Make your test study time the **last** thing you do that night before preparing for sleep. Do not clutter your mind by reading, talking on the phone, or watching television. Research shows that the last thing on your mind before sleep will stay with you through your sleep cycle. If you have ever intensely worked on an idea or problem right before going to bed, you may have experienced waking up in the middle of the night with the solution or a new insight. Your mind, in a relaxed state, can keep working while you are sleeping.

- The **morning of the test, pay attention to your nutritional needs**. You need to keep your energy level constant over time. Eating the wrong foods can make your energy too high, too low, or short-lived. Avoid caffeine (coffee, tea, soft drinks, chocolate), which can upset your stomach and make you nervous. (If you need help waking up, remember that light exercise will stimulate you mentally and relax you physically.) Avoid sugary foods such as donuts, some cereals, and coffee cakes. These foods give a rush of energy right after they are eaten, but the energy "peaks" quickly and then "crashes". Avoid greasy foods such as fried eggs or hash browns. Grease is hard to digest, will cause sluggishness, and can upset your stomach. Avoid acidic foods such as orange, tomato, or grapefruit juice. If you are nervous, you will already have enough acid in your stomach.

  Do not eat too much, because you will get sleepy. When you eat a big meal, digestion becomes complex. More blood is channeled to your stomach, which leaves less blood in your brain.

  What's left to eat!? There's plenty, really. Eat a light breakfast, such as fruit (cantaloupe, banana, or apple is fine) and a piece of lightly buttered or dry toast. A healthy (less sugar) cereal is fine, with a glass of milk. A snack, such as a lean meat sandwich, or eggs (prepared in very little oil or butter), also makes a good pre-test breakfast. Protein stays with you over time, decreasing the chances of irritability from hunger, and it will help stabilize your blood sugar so that your energy level will be more even and longer lasting.

- The morning of the test or right before the test, **don't be tempted to take a hasty look at your books or notes**. Remember, you reached a point the night before where you felt comfortable with the material.

Why risk losing that feeling by a rushed (and probably confusing) look at the material you have already gone over?

- **Ignore what other students have to say right before a test.** Often they are confused and frantically searching for solutions at the last minute. Most of the time they will be wrong anyway. And listening to snatches of conversation about problems taken out of context is extremely confusing. Keep your mind clear, and remember, in your review the night before, you felt good about what you know. Do not allow yourself to be confused by pre-test chaos.

DURING THE TEST

- Remember to **sit at the front of the room**. By doing so, you will be less distracted by what the other students are doing and you will be able to concentrate more on what **you** need to do.

- Carefully **read the instructions** before beginning any problem.

- **Go to a problem that is familiar** to you. You don't always have to work the problems in sequence. So begin with one that you are confident you can work correctly. You will be less likely to engage in negative self-talk. Fears of failing will not be as likely to enter your mind.

  And by first working the problems you know best, you will make positive progress toward completing the test. Knowing that you are making progress, you will feel more confident and relaxed. If you get stuck on any problem, do not spend too much time trying to work it. Work the problems you know first, then return to the difficult ones later. Your concentration will be much better.

- Focus all your attention on the problem you're working. **Don't let your feelings about a prior problem interfere with your performance on the current one.** Keep your emotions under control. Anger or frustration will get in the way of what you **can** do. For example, tennis professionals are often faced with a bad play or call that can interfere with their ability to concentrate. Consistent winners ignore such distractions and concentrate on each individual play. Remember, each problem is a separate chance to score points.

- **Show all of your work!** It is impossible to get partial credit for a problem when you have only written the answer and it is wrong. Most math problems require several steps of work to get to the answer. When you take shortcuts or try to do too much in your head, you will be more likely to make careless errors.

- **Watch out for careless errors!** Remember, the best attack is a good defense. Professional athletes say one of the secrets to success is avoiding mistakes.

  Far too many students could have scored 10 to 20 points higher on a test if they simply would have been careful. Even when you get partial credit for a problem, those one to two points off for each careless error add up quickly and, in some cases, will be the difference between passing or failing, or making an A instead of a B. When you work the steps to a problem, **double check each step before you go to the next.** In this way you will eliminate careless errors before working too far into the problem. It is easier to catch mistakes right away than to try to find them later.

  Another way to catch careless errors is by recognizing when a step or answer doesn't make sense. For instance, if you are working a problem which requires you to multiply 8.14 x 7.974, you would expect the answer to be about 8 x 8 or 64 because both 8.14 and 7.974 are close in value to 8. So if you get an answer of 649.0836, you should know that you have made the careless mistake of misplacing the decimal point. **Mathematics involves common sense.** Take advantage of your own basic knowledge when working math problems.

- When taking a **multiple choice** test, read carefully for what the problem is asking. Then read and think through **all** the choices. Whoever made up the test has probably thought of all the wrong ways to read and answer the question, so those choices will be there, too. Thus, what appears to be the most obvious is not always correct. **Do not select the first familiar answer**. You are avoiding math when you move too quickly to the next problem.

- **Watch out for negative self-talk and anxiety**. Do not let negative self-talk fool you into thinking you cannot do something before you have had the chance to give your best effort. Do not be afraid to try. Trying helps you learn, even if you do "fail." And remember, you are only taking a test; you are not being physically threatened, and you are not going "to die." Any test is just one of many that you will have to face in life. The more you take them, the more you will get used to them. And the more you control your negative self-talk and anxiety, the better you will do.

AFTER THE TEST

- **Congratulate yourself** on having worked hard. Do not pay much attention to what other students say after a test. Wait and see for yourself how **you** did.

- When you get your test back, make sure you **know and understand your mistakes. Mistakes are an opportunity to learn**. If you can, go over the test with your teacher or tutor. Recopy the problems you missed and try them again. If you are not allowed to look at or keep your test, ask your teacher to tell you what **types** of problems you missed. Know exactly what pages you should review again and what problems you should practice to help clear up any trouble you may still be having.

- If you do not earn the grade you hoped for, remember that **trying is as important as succeeding**. Success might not always come as quickly as you want, but you must not stop trying. Look at the outcome in positive rather than negative terms. For instance, the problems that you got right mean that you are capable of learning and understanding, and you can work toward getting more problems right the next time.

CHAPTER 5

What Every Parent Can Do To Help

Your involvement is the key to your child doing well in school. Students excel when parents take an interest in the learning process. Visiting with your child's teachers and helping with homework is a good start. But you can help your children have positive experiences with math in numerous other ways.

• **Speak no evil!** You are the most important role model for your child. Even small children are aware of how you feel and what you say. Children easily absorb attitudes, including any negative feelings you may have about math. Do not transfer your fear or dislike to your child. Do not groan or make faces when the subject is brought up. Do not make comments about how much you *"can't stand math"* or *"don't see the use for most of it."* If you are afraid of math, admit it and discuss with your child the origins of your fear and what can be done about it. And if he or she voices negative feelings, help your child understand from where these feelings come and what they mean. Do not let **your feelings** be your child's excuse for avoiding math.

- **Let your child know that you *expect* him or her to succeed in math.** Remember the study cited at the beginning of this book. An important reason Asian students do so well in math is because their parents have positive attitudes about the subject, and they **expect** their children to want to learn it. And other studies have shown a child's confidence in his or her math ability is more directly related to what the parents think about their child's competence than to the child's actual achievement record.

  High expectations also means wanting your child to make the best grade possible. You would not be happy if your child barely passed English or history. The same should be true of math. "Barely passing" is not enough. Students are capable of doing better. They may need help and encouragement, but they can do as well in math as in any other subject.

- **Help your child establish a routine for studying.** Besides ensuring that there is a quiet place where your child can work without interruptions, you should also set a time your child is apt to work best. For instance, one survey suggests that successful elementary school students work best immediately after school, whereas older school children need more time to unwind.

- **Monitor your child's progress in math**. When parents monitor home and class work, students complete more assignments, have higher test scores, and higher math grades. When talking to your child after school, go beyond open ended questions such as, *"How was school today?"* Also ask some specific questions (in your sweetest voice): *"What do you need to study tonight?"* or *"Do you have any tests this week?"* Try not to pressure your child, but do let him or her know that you care and want to help.

  If your child rarely has homework, meet with the teacher to see if assignments are completed in class or if there is material to be worked or studied at home. (High school teachers assign about 10 hours of homework a week. However, most students say they spend considerably less time than that on homework, and about 10 percent say they do none at all.)

  When meeting with the teacher, ask to see any work or graded papers. Check classroom participation. Get an evaluation from your child's teacher. **It is important to know as soon as possible whether or not your child may need extra help.** Let everyone know you are serious about your child doing the best he or she can in math.

- **Don't let your child skip material.** Children often become overly competitive in a math class. In grade school, they may race through their work so that they can move to the next higher book. Often their books are color-coded and there is a status symbol attached to the workbook in which the child is working. Help your child's teacher. Periodically, look through your child's workbook to check for blank sections. Get him or her to talk about math; make it a game where he or she explains some of the problems to you. Your involvement in your child's daily work will help prevent gaps in his or her mathematical knowledge.

- **If you do tutor your child, please be careful.** Don't rush. Try to schedule some unhurried time to work with your child. Remember that you are there to provide encouragement and guidance. And although it will sometimes be tempting to just provide the answers to the problems, it is important that you guide your child through the process of learning.

Be aware that helping your own child with homework can sometimes be frustrating. So if your family tutoring sessions result in shouting matches, consider getting help outside the home. You will not be the first parent to find difficulty in teaching your own child.

- **Praise your child's good work.** All of us know that praise and encouragement inspire us to do our best. Notice all of the good things your child is doing and give the credit he or she deserves.

- **Do not accept excuses for not succeeding in math.** Watch out for excuses such as:

"The teacher didn't explain it."

"We didn't know those problems were going to be on the test."

"I just can't understand it."

"The test wasn't anything like our homework problems."

"I'm never going to use any of this."

"I'm good in English, but I'll never be able to do math."

"The problems are so stupid."

"Hardly anyone got the problems right."

"My teacher hates me."

Be responsible. You and your child must do every-thing possible to ensure success in mathematics. So listen to his or her complaints and, **together**, get to the **source** of the trouble so that positive action can be taken. The above excuses are, of course, negative and passive statements that are, in a word, **excuses.**

- **Monitor your child's use of calculators to work homework problems.** Calculators can be helpful to the educational experience, but only after students have mastered basic skills. Some teachers report working with fifth graders who must still count on their fingers when adding, say, 9 plus 4. And many math anxious adults are so dependent on calculators that they become terrified of situations when they are not able to use them. Check with your child's teacher to know whether or not calculator use is appropriate for your child.

- **If your child changes schools, pay close attention to his or her progress in math, right from the start.** Different schools often have different math curriculums, and if your child is missing knowledge, ask your child's teacher what can be done to help your child catch up.

- **Make sure your child gets help when needed.** When your child expresses frustration or if the teacher recommends help outside of class, know your options. If you are unable to provide help yourself, ask about tutorial programs, private tutors, parent workshops, and homework hotlines. Finding the right help for your child will have a dramatic impact. He or she will begin to understand math. This understanding will lead to a better attitude, an increase in confidence, and greatly improved grades. Understanding and succeeding in math are accomplishments. Your child will feel good about himself or herself and what he or she is capable of doing.

  If your school does not offer a free tutorial program and you cannot afford a private tutor, **get creative.** Perhaps a relative or friend can help. Contact local social organizations such as the United Way for a referral. And remember, tutoring help is usually short term. It will be needed only until your child catches up and regains confidence.

- **Help your child learn to cope with frustration and disappointment.** Do not try to overprotect your child from frustration or failure by making life too easy. Instead of making excuses for why an endeavor was not successful, help your child learn from mistakes. Children who do not know that failure is a natural part of the learning process may try to work at only those tasks they do well. These children need to learn that taking risks and learning from mistakes can help them grow.

When your child makes a poorer grade than he or she would have liked, remind him or her that **setbacks are temporary.** If the errors are analyzed, he or she can profit from what is learned and do better next time.

- **Encourage activities and provide toys that promote solving problems and succeeding.** Too often toys and games are stereotyped as being for boys or for girls. Unfortunately little girls are bombarded with messages that they should want toys such as dolls and doll houses, toy kitchens, little shopping carts, horses with long manes and hairbrushes, and make-up sets. These toys encourage homemaking, decorating, shopping, primping, and dating.

Boys, on the other hand, are taught to want toys such as transformers, erector sets, model cars, computer games, and battle action dolls. These toys promote building and spatial skills, strategic planning, a sense of control, and a desire to win.

In addition, boys are more likely to be encouraged to participate in team sports. This gives them an important advantage. **Sports are an excellent way to de-**

velop practicing skills and the will to do your best.
With sports, one gains the experience of having to use
skills when games count. Nervousness is overcome by
concentrating on the game, and enjoyment of competition follows. Growth occurs as skills develop, and one
learns to try his or her hardest no matter what the outcome. The child sees that progress can be made, especially with good coaching and practice, and self-esteem increases. These experiences are an invaluable
preparation for studying math and taking tests. So,
encourage **all** of your children to engage in some
sporting or competitive activity.

Children are easily influenced by the media and their
peers. Television tells them what toys to buy, and
they *must* keep up with the kids next door. In fact, it
may seem impossible to steer them in a different direc-
tion without a good-sized battle. But **take control!**

See that your daughter also has toys and games that
develop skills and strategy. Find a sport she énjoys
and encourage her participation. By the same token,
make sure your son learns to enjoy cooperation as
well as competition. Studies show that students who
do best on tests are those who do not accept classic
sex-role stereotypes.

CHAPTER 6

For Teachers and Tutors
(and Curious Students)

Students dislike or fear math for many reasons. Some find it boring when they are asked to memorize what is already known. They rarely have the chance to experience math as a creative process. They tire of manipulating numbers. They see no connection between learning math and becoming a better person or understanding human nature.

Some become frightened when math suddenly changes from computations to the abstract problems of algebra. Some have had a negative personal experience in the classroom or at home. Perhaps no one expected them to do well in math or maybe they were ridiculed by classmates when they asked a question.

A good teacher can help students overcome these feelings. With the right attitude and good teaching methods, a teacher or tutor can have a strong, positive effect on students' feelings about math.

Be aware that your own attitudes toward mathematics and your preconceived ideas about students' capabilities for learning math affect performance. A study of future teachers reported that 40% had less than positive attitudes toward mathematics. In addition, these future teachers *expected* male students to do better in math than female students. And there are other inequities. For example, a study in Montana showed that teachers, when grading math papers, tended to give the benefit of the doubt to white students but not to American Indian students.

These studies, and others, reflect unhealthy attitudes that can dramatically undermine the potential that **all** students have for learning math. What you **believe** about your students is often a self-fulfilling prophecy.

Remember, there is no "type" of person who will fare better in math. Success is related to expectations, confidence in ability, and hard work. The belief that mathematical ability is greatest among white males is usually based on results of standardized tests (such as the SAT). Rather than comparing the innate ability of males to females or minorities to nonminorities, these test results usually compare people who have simply had more math to those who have had less. And more often than not, white males take more math courses.

Until high school, there are no real differences in mathematical performance between males and females. Later differences are a product of *choices*. Not unexpectedly, girls and minorities, not encouraged or expected to continue with mathematics, fail to enroll in as many high school math courses as do white males. A study of students in one California university reported that 57% of entering males, as opposed to only 8% of entering females, had four years of high school math.

Certain minorities have also avoided a math curriculum. For instance, another study from a California university reports that 79% of its Asian students and 72% of its white students were in a precalculus math sequence. The figures for Hispanic and Black students were 25% and 20%, respectively.

Your expectation that all students can succeed in math is critical in bringing about equity in math performance. And there are ways to help **all** students enjoy mathematics.

- **Get to know your students, their backgrounds, and their feelings with respect to math.** Students appreciate a teacher's interest in them. Knowing that their teacher cares will make students strive harder to learn the material and will make it easier for them to ask questions.

A good way to begin a math class at the beginning of the semester is to **get students to write about their personal histories with math.** Here are some suggested questions:

1. What was the last math class that you took? When did you take it? What grade did you make?

2. Is math important? Why? How do you use it? How will you use it later?

3. What are your past and present feelings about math?

4. Were there any significant experiences in your past that had a strong negative or positive effect on you and your feelings about math? Explain.

5. How do math tests affect you? How do you <u>feel</u> when taking a math test? What do you <u>think</u> about?

These questions can be written on the board and answered in class or handed out as a homework assignment. It takes students about 20 minutes to respond in writing. Besides giving you some groundwork with which to evaluate students' needs, it is a good opportunity for students to practice their writing skills.

After reviewing the responses, decide which ones to share and discuss with the class. Often, many students will have the same feelings or past experiences. It is a relief to them to know that they are not alone. Math anxious students are typically quiet. They are afraid to ask questions in class because "everyone will know how dumb they are" or they fear ridicule. You can bet that because of past bad experiences, they are going to hide their fear of math and lack of understanding for as long as they can, or until they know they can trust their teacher and their classmates and until they understand they are not alone. An open discussion of feelings will instill a helpful, supportive, and cooperative atmosphere that will help the whole class.

The above exercise will provide a springboard to tap into and understand feelings. It also provides an important cathartic release for students. They will learn, in general, that positive experiences with math lead to good feelings and negative experiences lead to bad ones. Sometimes these negative feelings have been harbored for **years.** Students are not even aware of them until they begin writing and find they have internalized negative consequences as personal failures. The resulting loss of self-esteem undermines their abilities to learn math.

This exercise gives you and your class the opportunity to discuss **positive steps** toward a successful learning experience. As a result, your students will be much better prepared to learn.

This exercise also lets students know that you care and are there to help. But first, you must be able to empathize with them. *"Of course you feel angry (or anxious),"* *"That must have upset you,"* or *"I understand how that would have disturbed you,"* are the kinds of things that students who fear or hate math need to hear. When you honestly accept their feelings, trust will grow. Then, as you get to know one another, there will be opportunities to delve further into the source of these feelings, and together you will be able to arrive at a responsible course of action.

You can also learn from the experience. Being aware of an individual's negative feelings does more than help teachers with a particular student. Understanding how past experiences influence students' attitudes toward math will provide valuable insights as to what works and doesn't work when teaching.

• **Practice patience and encouragement, not speed.** Most students learn better in a **relaxed** setting where speed is not emphasized. They need time to digest information and formulate questions and opinions. How quickly math is worked is not as important as taking time to understand. Let students know that persistence and patience are the keys to learning. Encourage their efforts to understand math, and let them know that even mathematicians must sometimes work for years to solve problems.

When helping a group of students, teachers and tutors should take care not to hurry from student to student. Instead of quickly finding a student's mistake and moving on, **sit down next to the student.** Take time to know if you need to explain concepts that precede the problem. Then discuss all parts of the problem and watch as they try another similar one. Invite other students working in the same lesson to work together.

A student reports, "*I remember in the fifth grade, the teacher I had was very fast in teaching. She only paid attention to the 'brighter' students and left the 'slow' ones behind. I was afraid of my teacher because she would always shout at the ones who were slow.*"

Another student says, "*My teacher had no patience with me and let me know it from the start. It made me feel dumb.*"

On the other hand, a student reports that a positive math experience occurred when, "*the teacher enjoyed taking the extra time out to really explain the problem.*"

And another student shares similar feelings by saying, "*I had always hated math until I finally had a teacher who made me feel at ease and more comfortable.*"

- **Encourage a cooperative and comfortable atmosphere.** Mathematics has a reputation for being very competitive. Students who work more slowly feel left behind, and sometimes students will race through their work without fully understanding what they are doing.

  Perhaps too much emphasis has been placed on "doing your own work". From time to time, have students work on problems or projects in small groups. Not only will students feel more involved with math by helping each other, they can also help each other with their feelings about mathematics.

- **Vary your teaching materials and show the usefulness of mathematics whenever possible.** A New England study of students in grades six through twelve said that students are often bored with math due to its lack of variety. Students feel that math is sterile and alienating.....something to be memorized.

Educators who address this problem suggest that teachers give creative problems that enhance the understanding of definitions and underlying concepts. Avoid emphasizing memorization or too many repetitive drills.

Give word problems whenever possible. In fact, have students suggest applications, then construct word problems from their comments and ideas. They will become more involved with the material and will see how math can help them better understand their world.

At every opportunity, point to newspaper articles that demonstrate how math is used daily. Have students find articles and write brief reports.

Another idea is to assign projects. Sports, politics, religion, cookie sales, or social issues.....find out what interests a student or a group of students. Meet with them and discuss ways in which mathematics can be applied to their selected interest. The finished product would include a written description of what they did along with mathematical computations. Students can then volunteer to share their projects with other classmates by giving a verbal report. Students will become more involved with mathematics and will be proud of their ability to apply it to an area of interest. By sharing their results, they will also see that math is widely used in all parts of life.

- **Show students how to read math textbooks.** Most students either ignore or, at best, skim the written material before each set of problems. If definitions are included in a test, students often memorize rather than try to understand them.

Explain that reading math is like reading instructions for "do-it-yourself" kits. Anyone who has ever assembled a toy from instructions knows that much patience is required. The temptation to scream or give up is tremendous. But once the instructions are slowly and carefully followed, they **can** be understood.

Help students understand that, even though mathematical definitions or theorems may sound confusing, they are, in reality, stating very basic ideas. Break apart a definition phrase by phrase. Explain that the individual pieces must be understood before the whole can be interpreted. Give examples that lead to the overall concept.

Then work in reverse to demonstrate how mathematical definitions are created. Introduce concrete problems that share a general mathematical characteristic. Have the students define the characteristic they have observed in general terms that can be applied to all problems of the same nature.

- **Treat your students with respect and sensitivity. Do not tolerate impolite behavior toward any student's efforts to learn.** Students who fear math are terrified of appearing foolish or stupid. They are afraid of asking or answering questions and of working problems on the board. A student said, "*I didn't understand just one part of the lesson, but the kids in my class still made fun of me.*"

Let students know you appreciate and respect their questions. Questions are an indication of a desire to understand. Even ill-formed or basic questions can present a creative opportunity to "reteach" material.

When a student asks about a concept already discussed in class, remember the quote by Emerson, "Each mind has its own method", and take his or her question as an opportunity to explain the problem in a new way.

Another math avoider reported,"*Working problems at the board really makes me nervous. It's the main reason I hate math.*"

Many students are terrified of working problems at the board. When you assign board problems, consider having small teams of students first work the problems together on paper. Then have one student from each team present the results on the board. Besides eliminating the usual anxiety associated with board work, you will have created a fun exercise that encourages cooperation.

• **Be sure to fully explain terminology or concepts.** A student tells us, "*In high school, the teacher used all kinds of mathematical-type words with more than three syllables. He would use college terms to explain algebra problems. I was really confused and afraid to ask any questions.*"

Until you get to know your students, be careful in your assumptions about what they already know or understand. A common pitfall of beginning mathematics teachers, especially those teaching foundations courses in college, is overestimating their students' mathematical knowledge. It never hurts to take the risk of "over-explaining", especially at the beginning of the semester. You will be surprised at the number of grateful students. Often these students are the quietest in class. They are afraid to ask questions, and sometimes a teacher will assume silence means understanding.

Another student relates, "*A lot of math teachers treat you as if you already know the material. I was behind from the first day.*"

And another student says, "*The teacher would work the problems so fast that it was all we could do to copy them from the board before they were erased. We never had time to understand.*"

When a teacher or tutor works problems rapidly, skips steps, and uses math terminology without reminding students of the meaning, they risk increasing students' fears that mathematics is *magical*. Math anxious students become even more convinced that only special people can do math and that they will never be able to understand it.

- **Speak clearly**. Make sure your students can understand what you are saying. Do not explain too rapidly, especially in beginning or fundamental math courses. Many students have trouble enough understanding math, especially beginning algebra, without having to worry about clearly hearing the teacher.

  One student said, "*One of my instructors could not speak clearly, so I was totally turned off. I could not understand what he was saying and missed out on half of what he was teaching.*"

  Another student who avoids math said, "*When I was a freshman in high school, I had a foreign algebra teacher who didn't speak English well. She thought that if you didn't understand her, you were stupid. Well, I failed math my first year in high school.*"

  Obviously, if you are a foreign-born teacher, you may not yet speak English perfectly. But do be aware that what you say may not always be understood. A suggestion might be that you ask the students to let you know if they are not able to understand you.

- **Never use math as a punishment.** One of the reasons some students dislike math is because they think it is boring. To give pages of problems as a punishment will make working math both dull and infuriating.

  A student remembers, "*When I was in the second grade, we were noisy when the teacher had to leave the classroom for a few minutes. As punishment, we had to do **pages** of math problems. To this day, I have bad feelings about math.*"

- **Acknowledge the correct things a student has done.**
  Math anxious students need to be reminded that they
  can do the work. Often they see only their mistakes.
  When looking at their work or helping them find their
  errors, be sure to point out what they have done right
  and give them praise. Consistently remind them that
  they **can** work math.

- **Give credit for how a problem is worked.** Teachers
  who emphasize only correct answers and give no
  credit for the process of working a problem give mathematics a reputation for inflexibility and lack of creativity.

  Students need feedback on **how** they have worked
  problems before they can really understand mathematics. Stress that when your students do not get a correct answer, they shouldn't erase their work. Instead,
  encourage them to show you their work so you can
  explain what they've done right and what they've
  done wrong.

  And to inspire creativity, **show other ways of working
  the same problem** to help your students understand
  that different approaches can have the same result.
  One way to demonstrate the variety of approaches is
  to assign a problem and give credit for how many
  different ways it can be worked. This exercise would
  also be an excellent group assignment.

  This **flexibility** will encourage your students to begin
  working on math problems without the fear that there
  is only one correct way. Remember, math anxious students think that only *special* people are good in math
  and that these *special* people usually work problems
  very quickly. Teach your students how to **flounder
  constructively** and to understand that this process is

normal for **anyone** working on a new mathematical problem.

- **Include a history of math whenever possible in lectures or tutorial sessions.** Mathematics can easily be "humanized" by giving historical examples of how men and women arrived at mathematical concepts. Knowing about these people's lives enhances the lectures and personalizes mathematics.

  For example, a good introduction to graphing ordered pairs might include a summary of the life of Rene DesCartes. Explain how he applied nautical direc-tions.....North, South, East, West.....by mapping two mathematical variables on the x and y axes. By seeing how they "travelled" together, he was able to visualize how variables are related.

  In addition, allow students to relive history by giving "hands-on" exercises where students can discover for-mulas for themselves. For instance, talk about circles and how they are the collection of points equidistant from a point called the center. Explain what the radius and diameter of a circle are. Then give the students several examples of circles. Have them measure the circumference and the diameter of each. See if they can find the general relationship between a circle's di-ameter and circumference. Show them that, histori-cally, that is exactly how the formula for the circumfer-ence of a circle was derived.

- **Provide an introduction to each day's lesson.** Stu-dents can better appreciate and understand your lec-ture if you explain what you will teach and why they should learn it.

  A student said, *"The teacher would just start putting all these algebra problems on the board, but we* leads them to pick an incorrect response. Multiple choice tests do not allow for giving partial credit. Students are discouraged when they have worked to learn a concept and do not get partial credit for what they have done right. They feel the test "tricked" them instead of helped them. They do not have an opportunity to feel proud of what they do know.

When multiple choice tests are graded by machine, students do not have the benefit of the instructor's comments and will not understand **why** they missed problems. Tests should be a positive learning experience that can redirect students to the areas they do not completely understand.

In conclusion, students who will soon be taking tests such as the SAT need practice with multiple choice testing. However, when first learning material, multiple choice tests rarely provide a positive learning experience and should be used on a limited basis.

- **Maintain an enthusiastic attitude towards mathematics and each student's ability to learn.** Students often use the guise of indifference or boredom as an excuse for not wanting to learn math. In reality, they may be unsure of their own ability. Your enthusiasm is of utmost importance in preventing students from feeling apathetic.

Teaching or tutoring mathematics can involve many elements learned in sports. You must be a good coach to help students understand the basics. Provide drills, supervise and correct performance, and insist that they can do it even when they want to quit. Be an enthusiastic cheerleader, rooting them on to success and encouraging them to keep trying even if they fall behind. And finally, remind them that math, like sports, requires time and practice as well as understanding. For example, being a good football player requires more than just knowing the rules. Only with a lot of practice can a person develop and demonstrate a skill. Test time, like game time, makes people nervous. But the more one has practiced beforehand, the more likely one is to succeed.

CHAPTER 7

Afterward

This handbook was written to help people who hate or fear math. As educators, it has been our great pleasure to see students overcome their negative feelings about math and do well.

But all of these students had something in common. They **wanted** to excel. Good teachers may inspire, and good parents may encourage. But ultimately, the student has to make a commitment to change the way he or she feels about math.

How do you change the way you feel? Well, that's definitely a challenge. But there are some things to think about that can help. Most of these thoughts have been discussed earlier in this handbook, but for clarification, let's review:

1. **Discuss feelings.** In order to fix a problem, we must admit there is a problem. And open discussion about math anxiety is the best way for people to understand that they are not alone in their feelings.

2. **Math is an important part of the world around us.** Math is everywhere. It is part of music, art, history, and everyday life. It is impossible to have any understanding of the world without a basic understanding of math.

3. **Math teaches us how to think.** To live well, we must all learn to think in a concrete, logical manner. To live

fully, we must also learn to think in the abstract. Thus, the process of learning math is perhaps the **best** way to enhance the capabilities of our minds.

4. **Math is not for the "magical few".** Math is not an inherent talent. With interest and determination, anyone can be successful.

5. **Confronting problems and then working to solve them greatly improves our self-esteem.** Feeling good about ourselves is what all of us are ultimately striving for, and nothing feels better than conquering something that gives us difficulty. This takes a great deal of courage and the ability to rebound from many setbacks. But that is a natural part of the growing process. And when the work is done and we can look back at our success, nothing feels better.

The Turtle
and
The Goldfish

No. This is not an old fable. But think about the turtle. Have you ever seen a turtle get anywhere with its head tucked in its shell? In order to move, a turtle **must stick its neck out.** And if you are going to make any progress in your life, you have to do the same thing.

Think about a goldfish. Goldfish are amazing. In a little goldfish bowl, a goldfish will never get over a couple of inches long. But in a large pond, the same goldfish can grow to be several feet long. **The goldfish expands to the size of its world.** Your brain works the same way. With very little information, the world your brain lives in is quite small. But as you feed your brain more knowledge, the world in which your brain lives keeps growing and growing.

In fact, **no one knows** how large your world can be. And you will never know unless you make a commitment to learning. It is said, "You only live once." And wouldn't it be a shame if you tucked in your neck and decided to live in a little bowl for the rest of your life?

Well, it really would be. **So don't!**

About the Authors

Angela Sembera has many years experience teaching math at the university level. An expert in the psychological and social aspects of mathematics education, Ms. Sembera specializes in working with students who hate or fear math. She has coordinated college tutorial programs, designed math anxiety curricula for high schools and colleges, and is a frequent lecturer to educators on the subject of math avoidance.

Ms. Sembera graduated from the University of Houston with a Bachelor of Science degree in mathematics and earned a Master of Arts degree in sociology from Cornell University.

Michael Hovis is the Executive Director of Public Films Inc., an award-winning filmmaking company. Mr. Hovis is the writer/director of educational films used by thousands of high schools and colleges throughout the United States and Canada. He earned his Bachelor of Arts degree in English from the University of St. Thomas in Houston, Texas.

Mr. Hovis has produced films that address a wide variety of social issues.....teenage pregnancy, drug abuse, AIDS, prenatal care, art, cancer prevention, mental health, and the psychological aspects of education.

Mr. Hovis and Ms. Sembera collaborated on the production of the film, *MATH! A Four Letter Word*. Viewed by millions of students and math anxious adults, their response to this film was the inspiration for this book.

For More Information

The authors are determined to reach the widest possible audience with information about math anxiety.

To learn more about workshops, videos, or this book, write to:

MATH! A FOUR LETTER WORD
The Wimberley Press
P. O. Box 1689
Wimberley, Texas 78676
1-800-MATH 987

*never really understood what we were doing or why
we were learning them."*

- **Be careful of how you design tests and prepare
  students for tests.** Math anxious students feel the
  most anxious when taking a test. Much of their anxiety is generated by a fear of the unknown. You can
  decrease this anxiety by letting your students know
  what to expect on a test. Make sure they understand
  what pages to review and what problems to practice.
  Stress understanding the instructions when practicing
  problems and write the instructions on the test using
  the same or similar words. Be fair about how you test
  your students. Do not try to trick or surprise them. If
  you would like to see how far they can stretch their
  mathematical knowledge, do it with a bonus problem
  or an extra-credit assignment.

When designing tests, state the point value of each
problem. Make sure students can show their work
and give them partial credit for what they have done
right. Avoid giving too many multiple choice tests.
Math anxious students tend to feel even more anxious
when presented with this type test. Some of the **problems with multiple choice tests** are:

> Not yet feeling confident with their abilities, stu-
> dents tend to doubt themselves when presented
> with too many choices.

> Having a history of avoiding math, math-anxious
> students avoid looking at all the possible an-
> swers on a multiple choice test. Instead of re-
> viewing and contemplating all the options, they
> pick the first response that looks familiar.

> Students may know how to work a problem, but
> they will sometimes make a careless error, which